Bibliographic information published by the German National Library:

The German National Library lists this publication in the National Bibliography; detailed bibliographic data are available on the Internet at http://dnb.dnb.de .

Imprint:

Print and binding: Books on Demand GmbH, Norderstedt Germany
ISBN: 9783346066114

This book at GRIN:

https://www.grin.com/document/505909

Zakir Islam

Air Quality Situations in Latin America and the Caribbean

GRIN Verlag

Air Quality Situations in Latin America and The Caribbean- A Review

Md. Zakirul Islam*
Department of Dairy Science, Bangladesh Agricultural University, Mymensingh-2202, Bangladesh

ABSTRACT

Air quality in countries and cities within Latin America and the Caribbean has deteriorated due to the rapid increase of the urban population, use of vehicles, poor waste management, extended burning of biomass-based fuels, large-scale combustion including power plants and industrial boilers which are strongly associated with fine particle emissions. In this thesis, data was gathered from online official sites of national and international legal sources such as the WHO Global Health Observatory, UNEP, ECLAC as well as the United Nations. We have used two major air pollutants namely PM10 and PM2.5 those are linked to serious health problems such as irritated eyes, nose and throat worsening asthma, chronic bronchitis, heart attacks, and arrhythmias along with cardiovascular and respiratory diseases and ultimately reduction in life expectancy. According to previous reports the maximum PM10 value was recorded in Ciudad Juarez, Mexico (136 µg/m3). Besides that, La Paz (Bolivia), Nuevo Leon, Jalisco & Toluca (Mexico), Lima (Peru) these cities contain comparatively high PM10 in their air. The minimum PM10 was recorded in Salvador, Brazil (17 µg/m3). Whereas the highest PM2.5 was recorded in Chillan, Chile (53 µg/m3) and the lowest PM2.5 was recorded in Punta Arenas, Chile (5µg/m3). PM10 and PM2.5 of most of the LAC countries and cities were higher than WHO air quality standards. The level of knowledge of air pollution's impact on health is very much limited in most of the LAC region. For improving health and quality of life, the direct and indirect health effects of air pollutants should be increased among both decision makers and the general population. A strong air quality management system is necessary to establish in LAC to implement the WHO Air Quality Standards.

Key words: Air quality, Latin America, Carribean, PM10, PM2.5

Table of Contents

INTRODUCTION

Air pollution is the release of several pollutants into the air which is directly or indirectly harmful to the living beings (Mackenzie 2016). Around 91% of the total world's population lives in areas where air pollution exceeds WHO air quality guideline (WHO 2018). Air pollution has turned into a major global concern in numerous urban communities around the world as a result of developing industrialization and expanding urbanization. It harms crops, creatures, forests, and waterways. It also adds to the exhaustion of the ozone layer, which protects the earth from the sun's UV light.

Air is mainly composed of nitrogen (N), oxygen (O), Carbon Dioxide (CO2), sulfur dioxide (SO2), carbon monoxide (CO), ammonia (NH3), nitrogen oxide (NOx), particulate matter (PM), volatile organic compounds (VOC) and other gases (Karim et al. 2000; Hackstadt et al. 2014). Air pollution is caused by different types of pollutants present in the air and due to unfavorable weather conditions (Papanastasioua et al. 2015). These pollutants are including various short-lived climate pollutants, particulate matter, etc. Short-lived climate pollutants (SLCPs) remain in the atmosphere for a short time which has a warming effect on climate. The major SLCPs consist of black carbon (BC), tropospheric ozone (O3) and methane (CH4). The largest source of SCLPs in Latin America is Mexico City(Mexico) (UNEP 2015). Furthermore, hydrofluorocarbons (HFCs) are responsible for air pollution. It is widely used in refrigeration and insulating foam (Molina et al. 2009). BC is another pollutant that absorbs sunlight and also melts ice and snow which cause global warming (Anenberg et al. 2012). 12% of the world's total BC emissions are originated from LAC. Black carbon remains in the atmosphere for a relatively short time for only days or weeks compared to other greenhouse gases, such as CO2. Accordingly, reducing black carbon emissions allows benefits almost immediately.

Particulate matter is defined as a complex mixture of extremely small particles and liquid droplets in the air (Zhang et al. 2015). Two classes of PM are considered by the EPA to be most deleterious to human health: PM10, which includes particles equal to or smaller than 2.5 microns in diameter and PM2.5, which includes particles equal to or smaller than 2.5 microns in diameter. They exist either as solid or liquid droplet aerosols.PM can be in the form of dust or can be derived through incomplete combustion reactions from both natural and anthropogenic sources. Natural sources for PM include volcanoes and forest fires; however, the amount of PM of anthropogenic sources is increasing. Anthropogenic sources include coal-fired power plants, internal combustion engines and other industrial processes, particularly those that use fossil fuels (Harrison, 2010).

The air pollution in Latin America and the Caribbean countries and cities is quite high. As the majority percentage of the area of LAC region are covered by industrialized zone, consequently, urban communities of LAC are creating a huge amount of pollutants that are causing air pollution (Thongplang 2015).

Air pollution is considered by the extreme environmental risk to human health (WHO 2019). Microscopic pollutants in the air can break our body defenses. It can also penetrate deep into our respiratory and circulatory system, damaging our brain, heart and lungs (WHO, 2018). Adverse health effects can be caused by BC, such as asthma and other respiratory problems, low birth weights, heart attacks and other cardiovascular diseases, and lung cancer (Kulkarni et al. 2006). In urban areas, each year 3.7 million people die exposed to contaminated air. Approximately 80% of these deaths are caused by ischemic heart diseases and strokes; 14% to chronic obstructive pulmonary disease or acute lower respiratory infections; and 6% to lung cancer. Particulate matter is most health related air pollutant in LAC countries than any other pollutants (WHO 2016).

In LAC, the monitoring system for air quality is very poor. But now most of the LAC countries are trying to improve their environmental standards to build up a strong air quality management system and to upgrade their related institutions. Construction, cement factories, mines and quarries, agricultural, petrochemicals are the principal industries to monitor particulate matters (Thongplang 2015).

This thesis work aims to investigate the air quality situation in Latin America and the Caribbean by making a review of the present status of air quality. This outline contains an evaluation of the current status of air quality observing countries and cities of LAC and the present levels of the most critical pollutants contrasted with the WHO Air Quality Standards. This review will also focus on the health effects of human causing by air pollution and development strategies to reduce air pollution in LAC.

THEORETICAL BACKGROUND

Latin America and the Caribbean

LAC covers a region that stretches from the Northern Centre of Mexico toward the southern tip of South America, including the Caribbean. Latin America is comprised of 21 countries in the American continent. These countries are all located in the south of the USA-Mexico border which starts with Mexico extending through Central America and parts of the Caribbean. LAC is surrounded by the North America continent at the north side, the Antartic continent at the south side, the Atlantic Ocean at the east side and the Pacific Ocean on the west side. (figure 2.1)

Figure 2.1 Map of Latin America and the Carribean
Source: https://www.pinterest.com/pin/534098837028326317/

The total land area of Latin America is 20,139,378 km^2 (UN 2019) which is double than that of Europe, but smaller than of North America, Asia or Africa. Latin America includes all of the continental countries of the Americas from Mexico to Chile and Argentina, as well as adjacent seas (Canziani et al. 1998).

The current population of Latin America and the Caribbean is 654,465,247 million. But, the population of Latin Americans will increase to 838 million in 2050. Latin America and the Caribbean population is equivalent to 8.55% of the total world population. The population density in Latin America and the Caribbean is 32 per km^2. Yearly population development rates will diminish from 1.68%, in the period 1995- 2000, to an expected 0.51% in the period 2040-2050, as per the medium prospect of the United Nations (Basso et al 2001; UN 2018).

Climatic Condition

Latin America contains an extensive assortment of atmospheres because of its geographical location. The climatic condition ranges from cool, frigid high rises, with a portion of a couple of ice sheets still found in the tropics to calm and tropical atmosphere. The atmosphere of Latin America ranges from the hot and damp Amazon River bowl to the dry and desert-like states of northern Mexico and southern Chile. Rain forests, deserts, and savanna all are found in this area. Earthquakes, volcanoes, and tectonic movements are very common all over Latin America and the Caribbean. Glaciers have ceased dramatically in the past decades and many of them have lost completely (Williams et al. 1998).

The atmosphere of Latin America is heterogeneous with biological communities and increased human population density. Northeast trade winds originating from the Atlantic Ocean bring heavy seasonal rains in Guyana, Suriname, Northern-east part of Brazil and Venezuela. Therefore, the climate of these countries is tropical humid (tropical wet and tropical wet/dry). Southeast trade winds bring rain in Sao Paulo, Curitiba, Porto Alegre(Brazil) and Montevideo(Uruguay). The climate of these areas is temperate (moist all year, winter dry, summer dry). Seasonal winds bring rain in Ecuador and Colombia. Deserts and steppes are present in mainly Argentina and Chile. The arid and semiarid climate is present there. Patagonia is located in this region which is comprised of Andes mountains and deserts. Rains come from the west of the Andes in this areas. The Atacama is also a renowned desert situated in Chile. The Andes block winds off the Atlantic from this desert. Winds also run from west to east through the Tierra del Fuego which is situated in the lower part of Latin America. Ocean currents (Peru current) brings cold surface water and here the air is very dry. In Peru, El Nino brings warm water instead of cold every few years. The coldest climate is observed in high altitude regions like La Paz (Bolivia). (figure 2.2)

The geographies of South America's climate are controlled by three key factors. The first and most important is the subtropical high-pressure air masses over the South Pacific and South Atlantic oceans. Large scale patterns of wind circulation and the location of the rain-bearing intertropical convergence zone (ITCZ) are caused by this subtropical high pressure. The second, presence of cold ocean currents along the continent's western side. It affects both

METHODOLOGY

To study the air quality of LAC, mainly the amount of particulate matter (PM10 and PM2.5) were used from different countries and cities. Data available on air quality policies and regulations for PM10 and PM2.5 in 14 different countries of LAC were considered for the most recent years.

The review of literature were accumulated online from authorized national and international societies and relevant permissible sources such as the Web sites or relevant publications of World Health Organization (WHO), United Nations Environmental Program (UNEP), Economic Commission for Latin America and the Caribbean (ECLAC), United Nations, World Bank as well as the Clean Air Institute (CAI) were also used. Data were furthermore collected from official government sources through internet searches as well as appropriate local officials. (ECLAC 2014).

The compiled informations were concised in related tables on air quality standards, characteristics of monitoring networks, in addition, historical data concerning air pollutant. Every single table was presented and conferred in the appropriate section. Air quality map was used to monitor the existing situations and subsequently air quality data were generated spatially.

RESULTS AND DISCUSSION

Air Quality Standard in LAC

According to WHO Air Quality Guidelines 2005, the standard PM10 value was 20 μg/m^3 and the standard PM2.5 value was 10 μg/m^3. From 2010 to 2014, the annual mean PM10 and PM2.5 value of 17 LAC countries were listed in Table 4.1.

Table 4.1. Maximum concentration levels of the main components of air pollution—particulate matter (PM10 and PM2.5) allowed in countries of Latin America and the Caribbean compared to World Health Organization Air Quality Guidelines (WHO-AQG)

Standards by country	PM10 (μg/m^3)	PM2.5 (μg/m^3)
WHO	20	10
Argentina	50	15
Bolivia	50	10
Brazil	50	20
Chile	60	20
Colombia	50	25
Costa Rica	20	15
Ecuador	50	15
El Salvador	50	15
Guatemala	20	10
Honduras	60	35
Jamaica	50	15
Mexico	60	12
Panama	50	15
Paraguay	45	15
Peru	55	15
Uruguay	50	15
Venezuela	50	15

Sources: www.leychile.cl/Navegar?idNorma=1025202

Air quality Data in LAC

Fourteen countries and some cities in countries of Latin America and the Caribbean had different amounts of PM10 and PM2.5 in their air. In most of the cases, PM10 and PM2.5 measured in 2010-14 in the air of these 14 countries and cities under those countries were more than the PM10 and PM2.5 limit according to WHO Air Quality Guidelines 2005.

Figure 4.1 showed that, In 2010-14, Ciudad Juarez in Mexico had the highest value of PM10 around 136 µg/m^3 which is 87% higher than the WHO standards. This city was a highly industrialized city and the number of population of Ciudad Juarez was also very high. Some other cities of Mexico such as Nuevo Leon, Jalisco, Toluca consisted of PM10 88, 87 and 82 µg/m^3, those are 80, 79 and 78% higher, respectively in terms of PM10 standard according to WHO Air Quality Guidelines. The second highest amount of PM10 in Latin America and the Caribbean was presented in the air of Lima(Peru) which was 94 µg/m^3 which is 81% higher than the WHO standard. La Paz had also high PM10 value of 82 µg/m^3. Buenos Aires in Argentina had 26 µg/m^3 PM10, Salvador in Brazil had 17 µg/m^3 PM10 in their air. Among all these cities, the cleanest air was presented in Salvador because it had the lowest amount of PM10 in its air. The air of Curitiba in Brazil also contained 24 µg/m^3 PM10. The air of Campinas, Jundiai, San Jose, Montevideo, Callao, Vina del Mar had PM10 respectively 35,30,27,26,34 and 34 µg/m^3 in their air which were very close to the PM10 standards according to WHO Air Quality Guidelines.

According to figure 4.3, in 2010-14, Chillan in Chile had the highest PM2.5 value of 53 µg/m3 which is 82% higher than the WHO standard (10 µg/m3). This city was not so highly industrialized like Mexico City and other industrially developed LAC cities but PM2.5 content of Chillan was very high than other LAC cities. The air of San Salvador (El Salvador), Guatemala (Guatemala), Toluca (Mexico), Lima (Peru) also had PM2.5 42, 41, 39 and 38 µg/m3, respectively which were above the PM2.5 standards according to WHO Air Quality Guidelines. In case of PM2.5, the WHO standard was only 10 µg/m3 around the whole world including LAC according to the WHO Air Quality Guidelines 2005. Among LAC cities, Punta Arenas(Chile), Montevideo(Uruguay), Cuenca(Ecuador), Pasto(Colombia) had the lowest amount of PM2.5 in their air respectively 15, 5, 9 and 9 µg/m3 but their amount was also above the PM2.5 standards according to WHO Air Quality Guidelines.

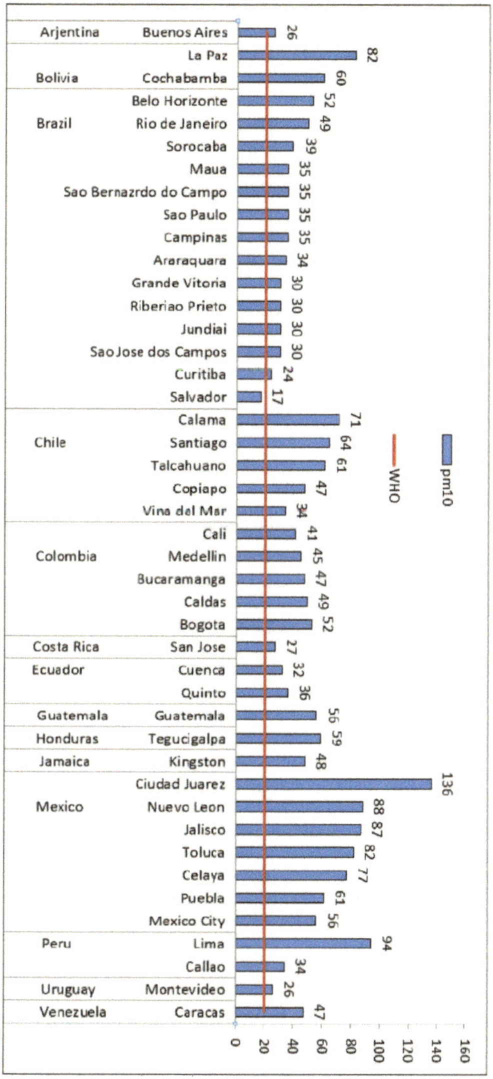

Figure 4.1: Level of Yearly PM10 (µg/m^3) Values in Cities of Latin America and the Caribbean (2010-14) Compared with the WHO Air Quality Guidelines, 2005.

Source: Collected from Rodríguez et al, (2016)

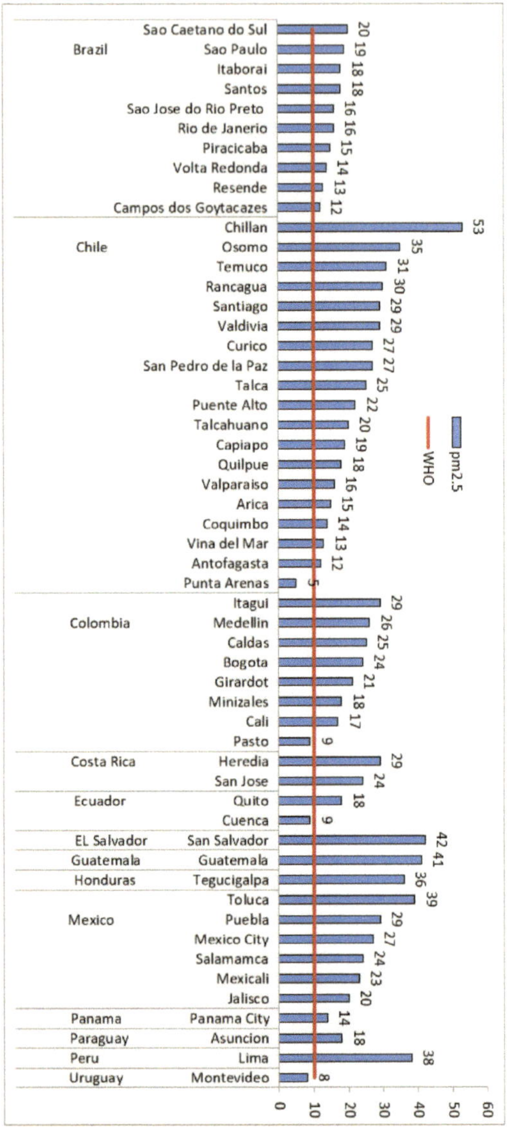

Figure 4.2: Level of Yearly PM2.5 (μg/m³) Values in Cities of Latin America and the Caribbean (2010-14) Compared with the WHO Air Quality Guidelines, 2005.

Source: Collected from Rodríguez et al, (2016)

Impact of Air Quality on Human Health

The world is getting hot and more crowded day by day. Our engines are also continuing to pump out dirty emissions and half of the world has no access to clean fuels or technologies (e.g. stoves, lamps). That's why the air we breathe is becoming very much polluted. Furthermore, nine out of ten people now breathe polluted air. Thus the people are suffering from various respiratory and heart diseases. Deaths rate due to pollutants present in the air is also increasing (WHO 2018).

Latin America and the Caribbean is one of the most urbanized regions of the world. So, its population is helpless against the impact of air pollution. In LAC, a large number of deaths occurs due to air pollution every year. This is very much alarming for ensuring a good healthy life in this region (Nowak et al. 2014).

In this region, major health problems caused by air pollution are acute respiratory disease, chronic disruptive pulmonary disease, ischemic heart disease, lung cancer, stroke, etc (Kim et al., 2015). In 14 LAC countries, the number of deaths caused by several diseases linked to air pollution was listed in table 4.2.

From table 4.2 it was analyzed that in 2012, among 14 LAC countries the highest number of people died in Brazil and the least number of people died in Costa Rica due to diseases caused by air pollution. In the other hand, the highest number of deaths occurred in Honduras and the least number of deaths occurred in Costa Rica caused by acute respiratory diseases. Due to Chronic obstructive pulmonary disease, the maximum number of deaths occurred in Mexico and a minimum number of deaths occurred in Jamaica (Gurjar et al., 2010). Stroke and lung cancer also caused a large number of deaths in LAC. In the case of both stroke and lung cancer, the highest number of deaths occurred in Brazil and the lowest number of deaths occurred in Costa Rica.

Table 4.2. Number of deaths attributable to ambient air pollution in 2012 in both sexes, by disease and LAC country

Country	Population (Million)	ALRI	COPD	Lung cancer	IHD	Stroke	Total
Argentina	4.4	85	192	1500	5256	2723	9756
Bolivia	11.0	265	68	88	1126	973	2521
Brazil	21.0	198	539	2901	12987	9616	26241
Chile	2.0	17	89	587	1087	1042	2822
Colombia	4.8	158	270	748	3688	1638	6502
Costa Rica	0.5	2	26	65	380	172	644
Ecuador	1.7	86	43	152	875	615	1771
Honduras	0.9	4088	532	985	7345	5486	18435
Jamaica	0.4	8	8	82	246	349	693
Mexico	12.7	467	742	1336	9983	4269	16798
Paraguay	0.7	42	16	108	529	465	1159
Peru	3.2	163	128	651	1949	1348	4239
Uruguay	0.4	4	14	169	308	217	713
Venezuela	3.2	125	126	814	3466	1582	6113

ALRI: Acute lower respiratory disease; COPD: Chronic obstructive pulmonary disease; IHD: Ischemic heart disease

Source: World Health Organization (2014).

Number of deaths by acute lower respiratory disease due to air pollutants in LAC countries against PM10 and PM2.5 of respective areas

In Latin America and the Caribbean countries, respiratory diseases are very common diseases linked to air pollution. As per figure 4.3.1, in 2012 death rate by acute lower respiratory disease increased largely in 14 LAC countries. In Honduras, yearly mean PM10 & PM2.5 were comparatively higher than other LAC countries and death rate by acute lower respiratory disease was also very high. Similarly, the number of death by acute lower respiratory disease was high in Mexico (467people) due to the presence of excess amount of PM10 and PM2.5 in its air. But in Costa Rica death rate by acute lower respiratory disease was comparatively lower than other LAC countries because of its lower yearly mean PM10 and PM2.5.

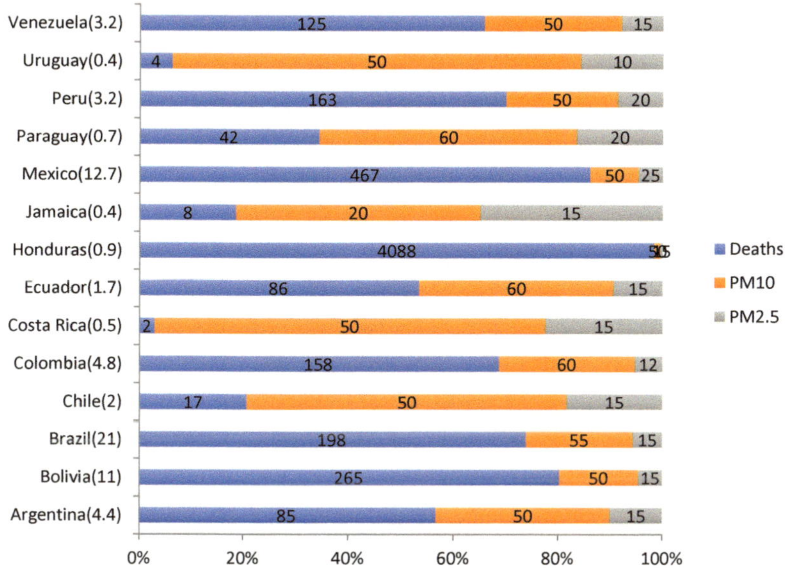

Figure 4.3.1: Number of deaths by acute lower respiratory disease due to air pollutants in LAC countries (million) against PM10 and PM2.5 of respective areas.

Number of deaths by chronic obstructive pulmonary disease due to air pollutants in LAC countries against PM10 and PM2.5 of respective areas

Likewise, figure 4.3.2 stated that in 2012 deaths by chronic obstructive pulmonary disease were comparatively higher in Mexico than other LAC countries because of higher yearly mean PM10 and PM2.5. In Brazil, the number of death by this disease was also high and alarming due to the presence of the high amount of particulate matter in its air. However in Costa Rica and Paraguay where yearly mean PM10 and PM2.5 were lower than other LAC countries, so death rate by chronic obstructive pulmonary disease was also lower in these countries.

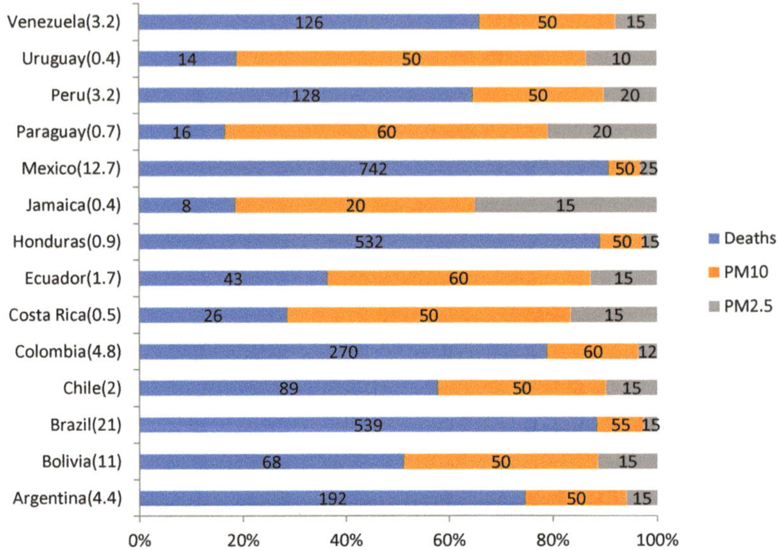

Figure 4.3.2: Number of deaths by chronic obstructive pulmonary disease due to air pollutants in LAC countries (million) against PM10 respective areas.

Number of deaths by lung cancer due to air pollutants in LAC countries against PM10 and PM2.5 of respective areas

Lung cancer is a very common disease in LAC countries and cities. Every year a large number of people dies by lung cancer in this region. Air pollutants like particulate matter (PM10 and PM2.5) is very much responsible for lung cancer and deaths due to this disease in this region.

According to figure 4.3.3, in 2012 the death rate by Lung Cancer was higher in Brazil because of it's higher yearly mean PM10 and PM2.5 than other LAC countries and the death rate by Lung Cancer was comparatively lower in Costa Rica because of its lower yearly mean PM10 and PM2.5. In Ecuador and Paraguay where yearly mean PM10 and PM2.5 were moderate, the death rate by Lung Cancer was also moderate there.

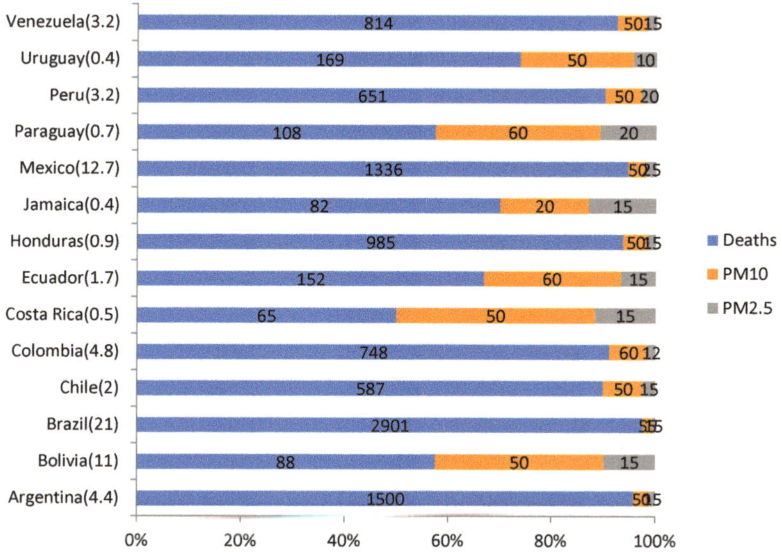

Figure 4.3.3: Number of deaths by lung cancer due to air pollutants in LAC Countries (million) against PM10 and PM2.5 of respective areas.

Number of deaths by ischemic heart disease due to air pollutants in LAC countries against PM10 and PM2.5 of respective areas

Figure 4.3.4 showed that, in 2012 death rate by ischemic heart disease in Honduras and Mexico was very high than other LAC countries. But in Costa Rica death rate by ischemic heart disease was comparatively lower than other LAC countries because of its lower yearly mean PM10 and PM2.5. So, it's clear that the death rate by ischemic heart disease was increased following the yearly mean PM10 and PM2.5.

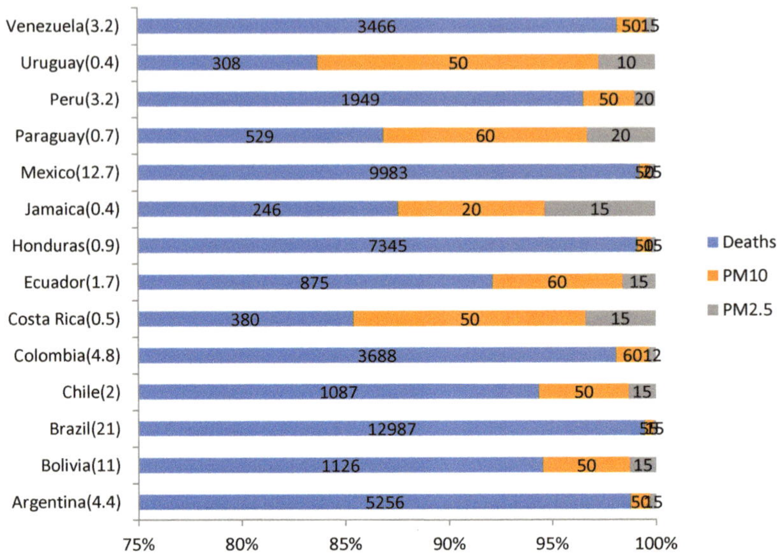

Figure 4.3.4: Number of deaths by ischemic heart disease due to air pollutants in LAC countries (million) against PM10 and PM2.5 of respective areas.

Number of deaths by stroke due to air pollutants in LAC countries against PM10 and PM2.5 of respective areas

Figure 4.3.5 analyzed mainly the number of deaths by stroke linked to particulate matters (PM10 and PM2.5) in LAC countries. In most of the cases, stroke is caused by the action of polluted air. Every year many people are affected by stroke and premature deaths are occurring all over the world. The death rate caused by stroke as per World Health Organization in 2012 was comparatively high in Brazil and Honduras, because of higher yearly mean PM10 & PM2.5 than other LAC countries and death rate by stroke in Costa Rica was comparatively lower than other LAC countries because of its lower yearly mean PM10 and PM2.5. The death rate by stroke was moderate in Peru and Colombia because of their moderate yearly mean PM10 and PM2.5.

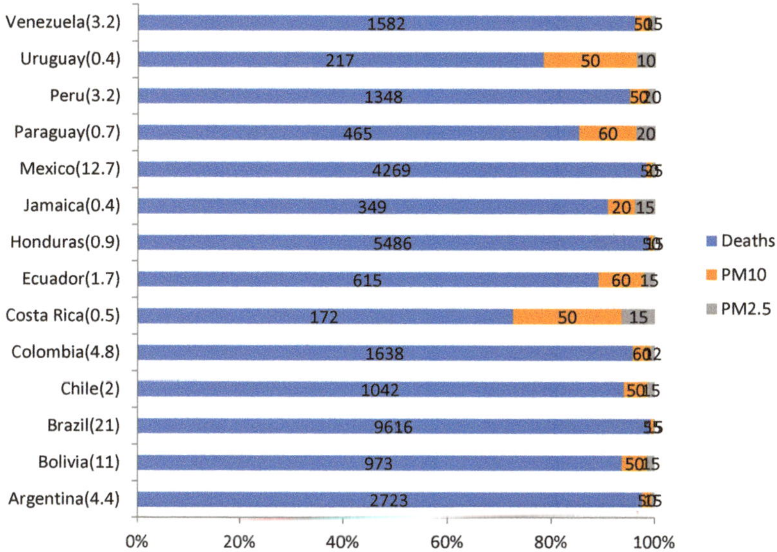

Figure 4.3.5: Number of deaths by stroke due to air pollutants in LAC countries (million) against PM10 and PM2.5 of respective areas.

Number of total deaths by different diseases due to air pollutants in LAC countries against PM10 and PM2.5 of respective areas

In 2012 total death rate by different diseases linked to air pollutants such as acute lower respiratory disease, chronic obstructive pulmonary disease, lung cancer, ischemic heart disease, stroke increased largely in LAC countries. Figure 4.3.6 showed that in Brazil and Honduras, yearly mean PM10 & PM2.5 were comparatively higher than other LAC countries and death rate due to air pollution was also very high. But in Costa Rica death rate due to air pollution was comparatively lower than other LAC countries because of its lower yearly mean PM10 and PM2.5. Therefore, it's easy to establish a relation between the death rate by different airborne diseases and the yearly mean PM10 and PM2.5 in different LAC countries.

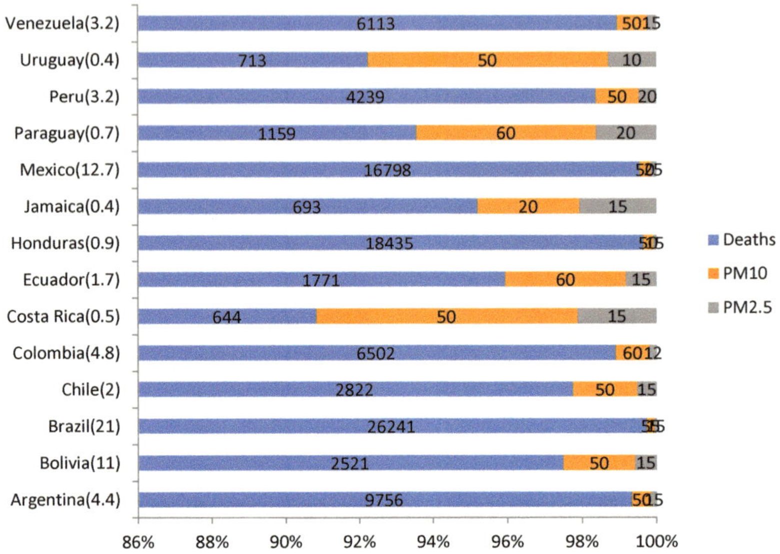

Figure 4.3.6: Number of total deaths by different diseases due to air pollutants in LAC countries (million) against PM10 and PM2.5 of respective areas.

Adaption strategies to improve air quality situation

Around 100 million people in LAC live in areas predisposed to air pollution. Generally, these areas are highly populated. In 2012, a total of 138,000 deaths in the Americas were attributed to ambient air pollution and household air pollution. Delegates, scholars, and non-administrative association of LAC are arranging different workshops and conferences to improve the air quality and environmental change issues that adversely affect air quality and the atmosphere. A regional expert Workshop on Air Quality was organized from 21 to 23 March 2018. In that workshop, 12 nations participated to make national strategies to address and reduce pollutants.

Following topics were the main agenda:

Current regional status of SLCPs: emissions, impacts, projections, policies and potential mitigation measures.

- ✓ Regional efforts on Air Quality Monitoring
- ✓ National strategies for improved air quality
- ✓ Key sectorial policies to control SLCPs: Transport and Household SLCPs within the fulfillment of the Agenda 2030 (UNEP 2018)

In 2018, WHO Global Conference on Air Pollution and Health was held at WHO headquarters in Geneva, Switzerland, participants of the conference considered the scientific evidence on air pollution and health and drew special attention to an immediate requirement should have taken to increase the global response to prevent diseases and deaths. This would help to achieve the goals in the 2030 Agenda for Sustainable Development related to Sustainable Development Goals 3. It is very important to reduce exposure to air pollution for protecting the health of children. More than half of all pneumonia deaths in children under five years of age are caused by air pollution. Besides, this early life exposure is connected with a maximized risk for many chronic diseases. Older people and individuals are affected with pre-existing cardiorespiratory conditions and diabetes are at notable risk. Eventually, the person exposed to increased levels of air pollution are at high risk and need to be protected by satisfactory measures. The need for a world free of air pollution was identified by participants at the conference. Additionally, an aspirational goal of decreasing the number of deaths from air pollution by two thirds by 2030 was highlighted. Every year significant global health care costs will be saved by reaching this goal (WHO 2018).

LAC nations have implemented the Regional Plan of Action on Atmospheric Pollution. The plan, which is the first of its kind in the world, recognizes the importance of the issue of air quality and encourages governments to identify the economic resources needed for the sustainability of the air quality monitoring networks. To reach the aspirational goal of decreasing the number of deaths from air pollution by two thirds by 2030, the following elements for a Geneva Action Agenda to reduce air pollution were put forward.

SUMMARY AND CONCLUSION

In recent times, air pollution has become a key issue in some countries of Latin America and the Caribbean because of urban development and growing industrialization. In accumulation to industrial processes often intensive in the cities, vehicle emission and fuel combustion are the principal sources of air pollution in LAC. Even though air-quality standards have been established in some Latin American countries but these are frequently exceeded. This review has provided the amount of two important pollutants PM10 and PM2.5 of different cities and countries in LAC. This has also provided an impact on human health due to poor air quality. The annual mean values of PM10 and PM2.5 in most measured countries and cities in LAC were significantly higher than the WHO-AQG-2014. Air pollution is responsible for huge costs by causing ill health and premature deaths in LAC. People are being affected by respiratory diseases, pulmonary diseases, heart diseases, etc.

As emissions of pollutants in LAC are high, yet there is a chance to scale up activities and accomplish enduring advantages for general wellbeing, atmosphere, and earth. More knowledge of pollution sources, atmospheric hazards & their impacts and effectiveness of remedial measures are needed. For improving health and quality of life, air quality education should be increased among all types of people. Another important priority is to increase awareness of the direct and indirect health effects of air pollutants among both decision makers and the general population. Establishment of a strong air Quality management system is very necessary for LAC to implement the WHO air quality guidelines. More investment should be made to improve the existing air monitory networks. It is very much necessary to increase the response to prevent diseases and deaths in LAC. Furthermore, access to clean energy and technologies in areas should be extended.

REFERENCES

Anenberg, S.C., Schwartz, J., Shindell, D., Amann, M., Faluvegi, G., Klimont, Z. et al., 2012. Global air quality and health co-benefits of mitigating near-term climate change through methane and black carbon emission controls. Environ Health Perspect, 120(6), 831–9.

Basso, E., Compagnucci, R., Fearnside, P., Magrin, G., Solman, S., Villamizar, A., Villers, L., Argenal, F., Artigas, C., Cabido, M. and Codignotto, J., LUIS JOSE MATA (VENEZUELA) AND MAX CAMPOS (COSTA RICA). Climate Change 2001: Impacts, Adaptation, and Vulnerability, p.693.

Berrueta, V., Edwards, R. and Masera, O.R., 2008. Energy performance of wood burning cook stoves in Michoacan, Mexico. Renew. Energy, 33(5) 859-70.

Bond, T.C., Doherty, S.J., Fahey, D.W., Forster, P.M., Berntsen, T., DeAngelo, B.J. and Kinne, S., 2013. Bounding the role of black carbon in the climate system: A scientific assessment. Journal of Geophysical Research: Atmospheres, 118(11), 5380-5552.

Brett, L., 2013. World regional geography, Available from: https://brettlucasgeography.files.wordpress.com/2013/03/world-geog-lecture-1-introduction.pdf. Accessed on 26 December 2018.

Canziani, O.F., Diaz, S., Calvo, E., Campos, M., Carcavallo, R., Cerri, C.C., Gay-Garcia, C., Mata, L.J., Saizar, A., Aceituno, P., Andressen, R., Barros, V., Cabido, M., Fuenzalida-Ponce, H., Funes, G., Galvão, C., Moreno, A.R., Vargas, W.M., Viglizao, E.F., and Zuviría, M., 1998. Latin America. In: The Regional Impacts of Climate Change: An Assessment of Vulnerability. Special Report of IPCC Working Group II. Intergovernmental Panel on Climate Change, Cambridge University Press, Cambridge, United Kingdom, and New York, NY, USA, pp. 187–230.

Carretier, S., Tolorza, V., Regard, V., Aguilar, G., Bermúdez, M.A., Martinod, J., Guyot, J.L., Hérail, G. and Riquelme, R., 2018. Review of erosion dynamics along the major NS climatic gradient in Chile and perspectives. Geomorphology, 300, pp.45-68.

Commission for Environmental Cooperation, 2011. North American Power Plant Air Emissions. CEC, Montreal, Canada, Available from: http://www3.cec.org/islandora/en/item/10236-north-americanpower-plant-air-emissions-en.pdf. Accessed on 15 December 2018.

Cifuentes, L.A., Krupnick, A.J., O'Ryan, R. and Toman, M., 2005. Urban air quality and human health in Latin America and the Caribbean. Inter-American Development Bank.

Economic Commission for Latin America and Caribbean, 2014. Statistical yearbook for Latin America and the Caribbean. Santiago, Chile: United Nations; 2014.

Gurjar, B.R., Jain, A., Sharma, A., Agarwal, A., Gupta, P., Nagpure, A.S. and Lelieveld, J., 2010. Human health risks in megacities due to air pollution. Atmospheric Environment, 44(36), pp.4606-4613.

Hackstadt, A.J. and Peng, R.D., 2014. A Bayesian multivariate receptor model for estimating source contributions to particulate matter pollution using national databases. Environmetrics, 25(7), 513-527.

Harrison, R.M., Jones, A.M., Gietl, J., Yin, J. and Green, D.C., 2012. Estimation of the contribution of Brake Dust, Tire Wear and Resuspension to non-exhaust traffic particles derived from atmospheric measurements. Environmental Science and Technology, (46) 6523-6529.

Heede, R. and Oreskes, N., 2016. Potential emissions of CO2 and methane from proved reserves of fossil fuels: an alternative analysis. Global Environmental Change, 36, 12-20.

Karim, M.M. and Ohno, T., 2000. Air quality planning and empirical model to evaluate SPM concentrations. Journal of environmental engineering, 126(12), 1116-1124.

Klimont, Z. et al., 2018. Technology-based global of black and organic carbon emissions from combustion. J. Geophys. Res, 109, 1-43.

Kim, K.H., Kabir, E. and Kabir, S., 2015. A review on the human health impact of airborne particulate matter. Environment international, 74, pp.136-143.

Kulkarni, N., Pierse, N., Rushton, L., Grigg, J., 2006. Carbon in airway macrophages and lung function in children. N Engl J Med, 355(1):21.

Latin America Map, 2019, Available from: https://www.maps.com/latin-america-wall-map.html. Accessed on 21 March 2019.

Martinez, M.A., Caballero, P., Carrillo, O., Mendoza, A., Mejia, G.M., 2012. Chemical characterization and factor analysis of PM2.5 in two sites of Monterrey, Mexico. J Air Waste Manag Assoc, 62(7):817–27.

Mackenzie, J., 2016, Available from: https://www.nrdc.org/stories/air-pollution-everything-you-need-know. Accessed on 12 May 2019.

Molina, M., Zaelke, D., Sarma, K.M., Andersen, S.O., Ramanathan, V. and Kaniaru, D., 2009. Reducing abrupt climate change risk using the Montreal Protocol and other regulatory actions to complement cuts in CO2 emissions. Proceedings of the National Academy of Sciences, 106(49), 20616-20621.

Nowak, D.J., Hirabayashi, S., Bodine, A. and Greenfield, E., 2014. Tree and forest effects on air quality and human health in the United States. Environmental pollution, 193, pp.119-129.

O'Ryan, R. & Larraguibel, L., 2000. Contaminacion del aire en Santiago: estado actual y soluciones . Centro de Economia Aplicada, Universidad de Chile, (75).

Papanastasioua, D.K., Melas, D. & Kambezidis, H.D., 2015. Air quality and thermal comfort levels under extreme hot weather. Atmospheric research, 152, 4-13.

Rodríguez, R.H., Soares, D.A., Texcalac-Sangrador, J.L., Moreno-Banda, G.L., 2016. Air pollution management and control in Latin America and the Caribbean: implications for climate change. Rev Panam Salud Publica, 40(3):150–59

Russell, M., Allen, D.T., 2004. Seasonal and spatial trends in primary and secondary organic carbon concentrations in southeast Texas. Atmos Environ, 38(20):3225–39.

Stodolska, M., Shinew, K.J., Acevedo, J.C. and Izenstark, D., 2011. Perceptions of urban parks as havens and contested terrains by Mexican-Americans in Chicago neighborhoods. Leisure Sciences, 33(2), pp.103-126.

Takahama, S., Russell, L.M., Shores, C.A., Marr, L.C., Zheng, J., Levy, M. et al., 2014. Diesel vehicle and urban burning contributions to black carbon concentrations and size distributions in Tijuana, Mexico, during the CalMex 2010 campaign. Atmos Environ, 88:341–52.

Thongplang, J., 2015. Particulate Matters: why monitor PM10 and PM2.5? Available from: https://www.aeroqual.com/particulate-matters-why-monitor-pm10-and-pm2-5. Accessed on 10 January 2019.

Troncoso, K., Castillo, A., Masera, O. and Merino, L., 2007. Social perceptions about technological innovation for fuelwood cooking: a case study in rural Mexico. Energy Policy, 35 (5), 2799-2810.

United Nations Environment Program, 2003. Progress in Phasing Out Lead in Gasoline, Governing Council of the United Nations Environment Program, Available from: https://www.iatp.org/news/countries-phase-out-leaded-gasoline. Accessed on 10 January 2019.

United Nations Environment Program, 2014. Regional Action Plan for Intergovernmental Cooperation on Air Pollution for Latin America and the Caribbean. Available from: https://www.ccacoalition.org/en/resources/regional-action-plan-intergovernmental-cooperation-air-pollution-latin-america-and Accessed on 10 January 2019.

United Nations Environment Program, 2015. World Meteorological Organization, Integrated assessment of black carbon and tropospheric ozone. Nairobi: UNEP, Available from: www.unep.org/dewa/ Portals/67/pdf/BlackCarbon_report.pdf. Accessed on 20 January 2019.

United Nations Environment Program, 2018. Integrated Assessment of Short-lived Climate Pollutants in Latin America and the Caribbean, Available from:http://ccacoalition.org/en/resources/integrated-assessment-short-lived-climate-pollutants-latin-america-and-caribbean. Accessed on 30 January 2019.

United Nations Environment Program, 2018. Meeting report - Regional Expert Workshop on Air Quality and Short-lived Climate Pollutants - Mexico - 21 March 2018, Available from: http://www.ccacoalition.org/en/node/2728. Accessed on 10 February 2019.

United Nations, 2018. Available from: http://www.worldometers.info/world-population/latin-america-and-the-caribbean-population/ Accessed on 10 February 2019.

United Nations, 2019. Available from: http://www.worldometers.info/world-population/latin-america-and-the-caribbean-population/ Accessed on 10 March 2019.

Upadhyay, N., Clements, A., Fraser, M., Herckes, P., 2011. Chemical speciation of PM2.5 and PM10 in south Phoenix, AZ, USA. J Air Waste Manag Assoc, 61(3):302–10.

United States Environmental Protection Agency, 2012. Report to Congress on Black Carbon, Department of the Interior, Environment and Related Agencies Act, 2010.

World Health Organization, 2014. Burden of disease from ambient air pollution for 2012: description of the method, source of the data and methods. Geneva: WHO; 2014, Available from: www.who.int/mediacentre/factsheets/fs313/en/ Accessed on 21 March 2019.

World Health Organization, 2016. Institute for Health Metrics and Evaluation. Global burden of disease 2015, Available from: http://www.healthdata.org/gbd. Accessed on 21 March 2019.

World Health Organization, 2018. Annual mean ambient PM2.5 (ug/m3) - from measurements, Available from: http://maps.who.int/airpollution/ Accessed on 21 March 2019.

World Health Organization, 2018. Global Conference on Air Pollution and Health, improving air quality, combating climate change – saving lives, Available from: https://www.who.int/airpollution/events/conference/en/ Accessed on 21 March 2019.

Williams, R.S. and Ferrigno, J.G., 1998. Satellite Image Atlas of Glaciers of the World: South America. U.S. Geological Survey, Prof. Pap.1386–I.

Winiwarter, W. and Klimont, Z., 2011. The role of N-gases (N2O, NOx, NH3) in cost-effective strategies to reduce greenhouse gas emissions and air pollution in Europe. Current Opinion in Environmental Sustainability, 3(5), pp.438-445.

Zavala, M., Barrera, H., Morante, J. and Molina, L.T., 2013. Analysis of model-based PM2. 5 emission factors for on-road mobile sources in Mexico. Atmósfera, 26(1), pp.109-124. World Health Organization, 2005. Air quality guidelines. Global update 2005. Copenhagen: WHO, 2006.

Zhang, R., Wang, G., Guo, S., Zamora, M.L., Ying, Q., Lin, Y., Wang, W., Hu, M. and Wang, Y., 2015. Formation of urban fine particulate matter. Chemical Reviews, 115(10), pp.3803-3855.